Once Upon a Time

Traveling Through Space and Time

"The distinction between past, present, and future, is only an illusion, however persistent."

Please give this book a positive review.

Every penny helps rescue and feed homeless and injured animals.

"A person with a negative tongue can give you the taste of his heart."

Introduction

Humans are motivated by curiosity and the need to know more. While our solar system is only about 4.5 billion years old, modern humans have only been around for about 200,000+ years. And physicists agree that in so many ways, we are just getting started with what we have learned about our universe.

What happens if you travel faster than the speed of light?

Special relativity states that nothing can go faster than the speed of light. If something were to exceed this limit, it would move backward in time, according to the theory.

So what is time?

While most humans think of time as a constant, physicist Einstein showed that time is truly an illusion; it is relative. Time can vary for different observers. It all depends on your speed through space. To Einstein, time is the "4th dimension."

Einstein's theory of special relativity says that time can slow down or speed up depending on how fast you move relative to something. And when approaching the speed of light, a person or an alien inside a spaceship would age far much slower than his twin at home. Furthermore, gravity can also bend time.

Albert Einstein was a German-born theoretical physicist who developed the theory of relativity, one of the two pillars of modern physics (alongside quantum mechanics). He is known to the general public for his mass energy equivalence formula $E = mc^2$, which has been dubbed "the world's most famous equation".

Einstein received the 1921 Nobel Prize in Physics "for his services to theoretical physics, and especially for his discovery of the law of the photoelectric effect", a pivotal step in the development of quantum theory. He subsequently realized that the principle of relativity could be extended to gravitational fields, and published a paper on general relativity in 1916 introducing his theory of gravitation. He continued to deal with problems of statistical mechanics and quantum theory, which led to his explanations of particle theory and the motion of molecules. Einstein also investigated the thermal properties of light and the quantum theory of radiation, the basis of laser, which laid the foundation of the photon theory of light.

In 1917, he applied the general theory of relativity to model the structure of the universe. Einstein moved to Switzerland in 1895 and renounced his German citizenship in 1896. After being stateless for more than five years, he acquired Swiss citizenship in 1901, which he kept for the rest of his life. Except for one year in Prague, he lived in Switzerland between 1895 and 1914.

In 1933, while Einstein was visiting the United States, Adolf Hitler came to power. Because of his Jewish background, Einstein did not return to Germany.

Einstein settled in the United States and became an American citizen in 1940. On the eve of World War II, he endorsed a letter to President Franklin D. Roosevelt alerting everyone to the potential development of extremely powerful bombs of a new type. He recommended that the US begin similar research. This eventually led to the Manhattan Project.

Once Upon a Time

The phrase, "Once Upon a Time," that we often hear at the beginning of every great fairy tale, is magical. But what is the true story of time itself? What do we truly know about time? Time seems to flow endlessly like water in a river. Once we were babies crawling on the floor, but so quickly we grew old, and needed a cane. But our past may truly not be gone at all, and our future may already be known.

Time flows in one direction: *toward the future*. However, some physicists have shown that much of what we believe about time is just an illusion. Time may not flow at all. Our past may still exists.

And our future may already be lived. Events that we often think can unfold in only one way, can also unfold in the opposite way, in the reverse direction. How could this be? And how could we be this much wrong about something as time so known and familiar? And if time is not what we think it is, then what is time? Did it have a beginning? And will it end eventually? Where did it come from? Why God (Allah) created time? It was created for a reason.

For example, the Holy Quran says: "It is God who made the sun a shining light and the moon a derived light, and determined for the moon phases so that humans may know the number of years and account of time."

"And God have made the night and day two signs. He erased the sign of the night and made the sign of the day visible so that humans may seek bounty from God and may know the number of years and the account of time. God has set out in detail for a reason."

We often would like to label and corner time as a thing, but it defines that completely by being momentary, by only having definitions that hearken back to the notion of time itself. We have been measuring time with ever-greater accuracy for many hundreds of years. The first clock was simple. It only ticked once a day, which only measured the rotating Earth. From the repetition of the Earth's daily rotation. On its axis to its yearly orbit around our sun.

We have always used the predictable and dependable motion of the Earth to measure time. We have always looked for and searched for things that repeat over and over again to measure time. That repetition, which cycle of things formed our clocks and time. That is all time becomes is some repetitive thing. Later we measured the Sun's motion with a sundial to divide the day into hours. The Earth rotates once a day, so we mark or tick off the days by watching the rising and the setting sun. Then with the swing of a pendulum, we divided hours into minutes and then seconds. And with the vibration of a quartz crystal, we improved accuracy to the thousandths of a second. The National Institute of Standards and Technology (NIST) measures true time. It is the U.S. official time.

At NIST they measure time with great accuracy using one of the smallest objects on Earth: an atom of rare metal called *Cesium*. Such atoms have a natural frequency. So anything that vibrates, that is giving you repetitive motion, can be a fantastic clock. The frequency at which the Cesium atom vibrates (ticks) is the official timekeeper for the world.

When that Cesium atom is blasted with energy, it vibrates (ticks) giving off pulses of light over 9 billion times per second. So NIST count the ticks of the Cesium atom. The Cesium atom ticks 9,192,631,770 every second. So each time you count up to 9,192,631,770, ticks, then one second has gone by.

For example, our normal watch loses or gains a second every few months. Here we are talking about a NIST clock that would only gain or lose a second in 100 million years. And this the kind of story, where we take one measure of time and we replace it with something that we decide is more accurate, has been the constant part of physics over hundreds of years.

However, no matter how accurate our clocks have become, time will always remain a mystery. A clock can tell you what time is it now, but they have not been able to tell us what time itself is? What is it that we are truly and actually measuring?

We may not know what time truly is, but the experience of the passage of time is an important and fundamental part of physics and life. We are always worrying about time, remembering the past, and making plans for tomorrow and the future, living life within time's ticks.

Now here is where our story of time begins. As the settlers moved westward after the War of 1812, the need for synchronized clocks became ever more critical. 3 types of time measurements were used: natural time, local time, and de facto railroad time. And time based on the natural movement of the sun throughout the day was still in use by individuals and was especially suited to a farming society.

The railroad station in Boston had a clock tower that helped travelers and citizens maintain a schedule that was synchronized to local time.

Clocks used synchronized astronomical time, which was based on time at the meridian of a specific location. It was displayed by town clocks and was very useful for town's government and to any citizen needing to synchronize their watch. Railroads ran on the time kept in the city where the line originated.

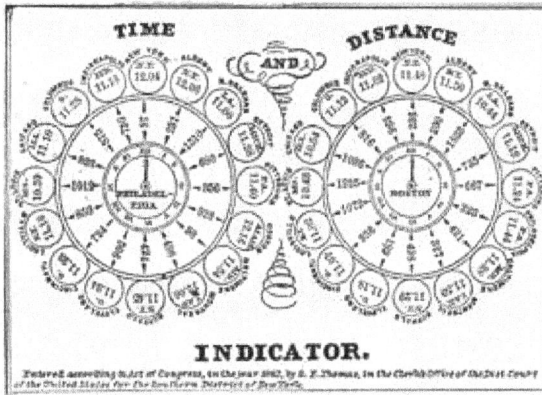

TIME AND DISTANCE INDICATOR.

By the 1860s, ascertaining the time for connections between train lines was complicated.

In the 1840s, New England railroads began publishing monthly schedules that they called timetables, in order to coordinate time between train lines. 80 different timetables were in use in the U.S. by the 1860s, making connections between train lines extremely difficult.

The British already addressed the issue of railway time, using the meridian of the Royal Observatory at Greenwich, hence Greenwich Mean Time. Because England was smaller, the problem was easier to resolve since it required only one time zone.

Scottish-born and Canadian engineer Sanford Fleming is credited with being the Father of Time Standards. Fleming spent his life working for the railroads. He was quite familiar with the issue of Railway Time. Fleming published his first notes on Terrestrial Time in 1878. Fleming recommended international standards as well, citing the situations of a steam ship traveler from England who upon arrival in North America transfers to a train.

A year later, Albert Einstein was born in Ulm, Germany. He was the first child born to Hermann and Pauline Einstein. Though he attended school as a young boy, he also received instruction at home on violin. By the age of 12 he had taught himself geometry.

At the age of 16 he failed an exam in order to qualify to study as an electrical engineer. But he stayed in school and decided to study math and physics to become a teacher. Einstein thought he would be good at this because he could think mathematically & abstractly while lacking imagination and practicality. In 1900 he graduated as a teacher of math and physics. In 1901 Einstein became a citizen of Switzerland.

His teachers did not think very highly of him, and so they did not recommended him for a job at a university. In 1901 he took a job as a high school teacher and married Mileva Maritsch. They had two sons prior to divorcing. Einstein later married his cousin Elsa Einstein.

In 1902 took a job at the patent office in Bern, Switzerland. It was a great place to see all of the great inventions of that time. The patents showed how to synchronize time with exchange of telegraph signals, to synchronize time by radio waves, all these patents made the synchronization of time, and what time was, and how it was measured, something immediately important and exciting for Einstein.

Years later, Einstein shook up science with his radical insight into the nature of time. The patent job strengthened Einstein's intellectual development and abilities. The job provided unexpected inspiration. Einstein realized that these attempts to synchronize time, are much more than simple creative inventions. Instead, Einstein quickly realized that they were revealing a deep crack in our understanding of time itself.

For example, time for Isaac Newton was something that was unchanging and regular. He said time always moves at the same rate. Time just goes along, and there is nothing we can do about it. Einstein said this was not correct. Time can run at different rates, which means that time for me may not be the same as the time for you.

Einstein meant that time is not just a static label on the whole universe; time is experienced individually. Everyone has their own private time, which runs at their own private rates. There is not time in a sense of a universal tick-tock; there are many different TIMES.

Einstein uncovered and discovered a hidden connection between time and space. What Einstein discovered was a link between motion though space and the passage of time. The more you have of one (space or time), the less you have of the other. To see how this exactly works, walk with me through space and time. For example, right now, I am heading due south at 3.1 mile per hour. And that means all my motion is in the southward direction.

But let me now turn onto a different road and head southeast. I am still walking at a speed of 3.1 miles an hour, but I am not making as much progress toward the south as I was before. And that is because some of my southward motion has been diverted, or shared with, my eastward motion.

Einstein realized that space and time are linked in much the same way that south and east are. And with this amazing insight, Einstein destroyed the common-sense idea that time ticks the same for everyone. Here is what he meant. Imagine a wolf sitting 300 feet away from you with his handler. If you are just looking at him, he would ignore you as long as you are not moving at all. But, you are moving in some sense.

You may not be moving through space towards the wolf, but you are moving through time. Your watch still keeps on ticking and ticking. And as long as you are standing still, not moving through space, Einstein said that all of your motion is through time. However, when you start to walk towards that wolf, because now you are in motion, the wolf's handler will perceive that your watch is ticking much slower. That is because from his perspective, some of your previous motion through time is now being diverted into your motion through space. And it is not just your watch. If we exaggerate the effect, he would perceive that all your movement, your voice, your feet, everything about you is slowing down. If you suddenly stop moving, the passage of time on your watch and his once again agrees.

This was Einstein's crucial insight: that motion through space will affect the passage of time. Why should your measurement of time depend on how you are moving? That does not make much sense. Time itself is running more slowly for the person who is moving. No one before Einstein has ever imagined that this sort of thing would truly ever happen. This is uniquely Einstein.

So why we never see such things in our everyday life? This is because at the very slow speeds that we move here on Earth, motion's impact on time is very tiny, so we just do not experience it. But the effect is real and can be measured. To measure it, you will need two atomic clocks and an airplane. This experiment was truly done in 1971.

Scientists flew an atomic clock around the world, and then they compared the clock's time to the one on the ground. As Einstein had said and predicted, the two clocks no longer agreed. The clocks differed by a few hundred billionths of a second, but that was enough and true proof of motion's effect on the passage of time.

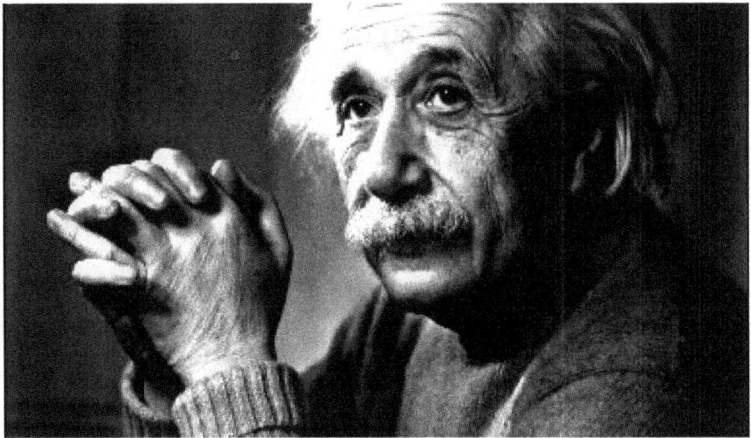

Space-time

Einstein's theory was tested again and again, and it all hangs tightly together. The theory forms the basis for the way we understand much of the way nature and time work. With the discovery of this unexpected link between space and time, Einstein said that the two could no longer be thought of as two separate things. Instead, space and time are fused together in what came to be called "space-time." So Einstein unified the idea of space with the idea of time into a 4-dimensional structure called space-time. And this fusion of space and time led Einstein to perhaps the most mind-bending realization of all: That difference that we all see between past, present, and future may only be an illusion.

In everyday life, we experience time as a continuous flow. But it is best to think of time as a series of moments or snapshots. Everything that happens in life can be thought of as the unfolding of moment, after moment, and after moment. If we picture all moments or snapshots lined up side by side, every moment here on Earth, every moment of Earth orbiting the sun, and every moment throughout the universe, we would see all events that has occurred, or will ever happen. Every moment in time, from the birth of the universe at the Big Bang, about 14 billion years ago, to the formation of stars in the Milky Way galaxy, to the creation of Earth 4.5 billion years ago, to the time of the amazing dinosaurs, to every event happening on Earth now, like reading this book right now.

You have to think about the seemingly simple concept of "now." For example, for me a list of things that I consider to be happening right now include the sound of the midnight tick on my watch, my dog just now chasing my cat, things happening far away, like an airplane flying to Jerusalem at this very moment, a meteor just now hitting Mars, and the explosion of a star at the far side of the universe. These and all other events that I might think are happening at the same moment in time, but in different regions of space, make up what I think of as "Now." You can picture them as lying on a single slice of space-time. Let me call it a "now slice." Common sense says that everyone will agree on what is happening or what exists, right now, moment after moment, after moment.

We would all agree on what lies on a given "now slice" of time. But Einstein showed that, when adding motion into account, this common-sense picture of time is no longer true. To understand what I mean, think of space-time as a loaf of bread. Einstein realized that just as there are many different ways to cut a loaf of bread into individual slices, there are different ways to slice up space-time into individual NOW slices. This is because Einstein said motion affects the passage of time, someone who is moving will have a different conception of what is happening right now, and so they will slice the loaf of time into a different now slice. Their slices will be at a different angle than yours. A person who is moving will tilt the knife, and will be carving out their slices at different angle.

They will not be parallel to your slices of time. To get a feel for the bizarre effect this can have, imagine a space alien, in a galaxy a billion light-years from Earth. And way over there on Earth, you at your desk. Now, if the alien and you are standing still, not moving in relation to one other, the alien clock and yours, will tick off time at the same rate, and so you and the alien share the same now slices, which cut straight across the loaf of bread (time). But what happens if the alien jumps on his flying machine and moves away from Earth and you. Since motion slows the passage of time (as shown in the atomic clock test), your clock and the alien's will no longer tick off at the same rate. And so if your clock and the alien's no longer agree, the now slices will no longer agree either.

The alien's now slice cuts through the loaf of time differently. It is angled towards the past behind you. Since the alien is moving at a slow, leisurely pace, his slice is angled to the past by only a minuscule amount. But across 1 billion light years from Earth, that tiny angle results in a huge difference (angle) in time. So what the alien would find on his angled now slice, what he considers as happening right now on Earth, no longer includes you or even 50 years earlier, before you were born. The alien's now slice has swept back through 100 years of Earth history and now includes events that we consider part of the distant past, like World War I. We can have actually tremendous disagreements on our labeling of now, what happens at the same time, if we are spread out far enough in space.

And if that is not strange enough, the direction the alien move makes a big difference too. For example, what if alien turns his space ship around and flies towards the Earth. The alien's new now slice is angled towards the future, and so it includes events that will not happen on Earth for 100 years, perhaps your great-great grandson teleporting from Jerusalem to Paris. Once we know that your now can be what I consider the past, or your now can be what I consider the future, and your now is every bit as valid as my now, then we are certain that the past must be real. And the future must be real. They could be your now. So this means past, present, future: all equally real. They all exist and are valid in time. So could Marty McFly could set out to fix the past or the future?

If you believe the laws of physics, then there is just as much true reality to the future and the past as there is to this present moment. That means the past is not gone, and the future is valid. The past, the future, and the present are all existing in exactly the same way.

Just as we think of all space as being "out there," we should think of all time as being "out there," too. Everything that has ever happened, or will ever happen. It all must exist in time. From the time Leonardo da Vinci painted the final brushstrokes on the Mona Lisa, to the Wright brothers' first flight, to your fist day at school, to events that from our perspective are yet to happen, like the first humans flying and landing on Mars. With this amazing and bold insight, Einstein shattered one of the most basic concepts of how we think of and experience time. Einstein said:

"The distinction between past, present, and future, is only an illusion, however persistent."

But if it is true that every moment in time already exists, then how do we explain the very real fact and feeling that time, like water in a river, seems to endlessly rush forward?

Maye we have been deceived and time does not flow at all. Perhaps the river of time is more like a solid, frozen river with every moment forever frozen and locked in place. The most vivid example the way the world is has to do with the flow of time. Physics does radical violence to this everyday experience of time. Our entire experience of time is always in the present. And all we ever aware of is that instant moment. There is nothing in the laws of physics that selects one now over any other now.

And it is just from subjective viewpoints that feels like things are changing. Just the way an entire movie can exists on film, all of time may already exist too. The difference is that in the movies, a projector lights up and selects each movie frame as it goes by. But in the laws of physics, there is no evidence of anything like a movie projector light that selects one moment over another. Our brains may create this impression, but in reality, what we all experience as the flow of time may be nothing more than an illusion. But if time is like a frozen river, does not flow at all, and all of time is "out there," is it possible to travel to the future or the past?

Traveling through space and time

If we could time travel, would it be anything like what we have dreamt of and imagined?

No one outside Hollywood has made a working time machine. But surprisingly, time travel is truly possible. One way to travel through time is to make use of a strange feature of gravity.

That familiar gravity force that keeps our feet planted to the earth, can have a profound impact on time. So how can gravity be used to make a time traveling machine?

Einstein's theories show that gravity, like motion, can truly affect time. It is as if gravity grabs and pulls on time so tightly, slowing its passage. Moreover, the stronger the gravitational pull, the gravitational force, the more time slows. Here on Earth, the effect is too small to feel and notice, but it is still very real. Compared to someone living on the top floor of a building, someone living on the bottom floor experiences time elapsing a tiny slower, since the gravity is just a tiny bit stronger closer to the ground.

Time traveling through a black hole

But if we could travel to a black hole, the effects of gravity on time would be truly huge. Formed when large stars collapse in on themselves, black holes have immense gravitational pull and force, billions of times stronger than the Earth's gravity.

If someone can see you traveling close to a black hole, they would see time for you slowing down so dramatically. You near that black hole will appear to your friends, (if possible to see you) back on earth, to be moving and talking very slowly, and biologically aging slowly. To them years are passing, while for you it might just be minutes.

So depending on the black hole's size and how close you get to it, if you spend an hour in near it, something like 100 years would have gone by back on Earth. You will have traveled to Earth's future. So when you return, you will find yourself in the future. Everyone else will have aged 100 years or already gone, but you will have aged only a couple of hours.

Time traveling through a Wormhole

Time travel to the future is one thing, but what about time travel to the past? That might be also possible too, using a thing that was predicted by Einstein's equations known as a *wormhole*.

If wormhole exists, it would be like a shortcut through space-time.

Tunnels in space that would link not just one place with another, but also one moment of time to another. A wormhole would connect one part in space-time to another space-time, which is at an earlier time, like a sort of subway system through time.

Let us say you wanted to go back in time to meet yourself when you began to read this book. If can find a wormhole that is connected here and there, then all you need to do is to step through it. It would be kind of weird to meeting yourself, but the real problem with time travel to the past is that things would get very confusing fast. For example, imagine you were to change something about your past, like preventing your parents from meeting.

That means you would never be born? So if you travel to the past, you cannot change things that we know are true about the present and the past because they already happened. So if you go back in time and kill who you thought was your grandfather that must have been some other man you thought was your grandfather, and everything must somehow become beautifully self-consistent, even if it is in a twisted way.

If we can travel to the past, why have we not been overrun by tourists from the future? Think about it! We have not seen intrepid time travelers popping into and out of our world. So it is safe to assume that time travel to the past just is not possible at least not yet.

But since Einstein's math has not yet ruled it out, we cannot dismiss time travel to the past entirely. So it is not all clear that it could ever be a practical reality, but at least in principle and in mathematics, it does not seem to be impossible.

While it seems for now that traveling to the past is out of reach, what about the fact, so common to our life experiences that time itself seems to move in only one direction, toward the future? We call this the *arrow of time*.

The Arrow of Time

The arrow of time is probably the most thing about the universe we live in that we do not completely understand well. Why we live in a universe that had a directionality to time is a big mystery. This is not true of space. In space, you can go from Jerusalem to Paris and then you can change your mind and go from Paris to Jerusalem. So there is a one-way aspect to time that we do not understand very well at a fundamental level.

Why time cannot go backwards? What does it even mean when we say that time goes forward from the past into the future?

Where the arrow of time comes from? Why do we only see events unfold in only one direction? Why we do not see them happen in reverse direction?

It must be the laws of physics. Because surely the laws of physics do not allow a bullet from a gun to move backwards. Actually they do. The laws of physics are the mathematical equations that we use to describe almost everything, from the behavior of atoms to the swirl of galaxies. They have been devised, and proved, and confirmed through many centuries of observation, and tests, and experiments. But amazingly, there is nothing in the laws of physics that says that events have to unfold through the familiar order we call "*forward in time*."

According to Einstein's mathematical equations, events could just as well unfold in reverse direction. Because all of the equations that we often use to describe what we see in the universe around us do not have an arrow of time attached to them. They are equations that work equally extremely very well moving forward in time, and moving backwards in time.

There is some contradiction between the physics, which sees fundamentally reversible, and so much of our life that seems very irreversible. Though it flies in the face of everyday experience, the laws of physics do actually say bizarre things like flowers growing back to seedlings, but how could this be?

The answer is not as far-fetched as one might think. Here is a reason. We all know what will happen if a child drop a glass of water. Now, the idea that the glass and water mess on the floor could somehow magically reverse itself and form back into a solid glass filled with water seems very absurd. But according to the laws of physics, this can happen. You just need to reverse the velocities of everything that is on the floor. Every piece of glass, every drop of water, every molecule, and every atom in the glass, water, table, and air. Just reverse all their velocities, and magic can truly happen. The glass of water is back in your hand. So if the laws of physics do not care about if the glass shatter or un-shatter, why we do not ever see such things un-shatter?

How can we square the laws of science and physics with everyday experiences?

Something must be truly missing in our understanding of physics, mathematics and the natural world. But what is it?

What is truly responsible for the arrow of time? Like many good mysteries, this one leads us to a graveyard and a grave in our search for clues and answers.

Entropy

In Vienna, near the final resting places of Brahms, Beethoven, Strauss, and Schubert, is a forgotten, 19th-century Austrian physicist Ludwig Boltzmann's tombstone.

Etched on top of his tombstone is a very elegant equation $S = k \log W$.

It is the mathematical formulation of a powerful concept known as *entropy*.

Entropy is a measure of something that everyone is familiar with: *disorder*, or *randomness*.

And it is a very important idea because there is a tendency of everything in our world and in the universe to move from order to disorder.

Here is a way to get a feel for such an idea. Take any book. All pages of it. The book is very ordered, with the first page followed by the second page, followed by the third, followed by the fourth page, followed by the fifth page, and so on.

Now tear all the pages out, and throw them in the air, and let entropy do its job. The pages will fly in the air and they will become very disordered. And the reason for this very simple:

There is only one way for the pages to land in order, 1, 2, 3, 4, 5,…, but there is a huge, many number of ways for them to land out of order, and so it is a billion more times and likely they will land in a total mess.

This is what we experience in life: things move from order to disorder always. In our book example, from a neat, ordered book to pages that are scattered on the ground in random order.

Everywhere we look, we see so many examples of entropy or disorder, and it increases with the passage of time. For example, an egg breaks and splatters on the ground. Ice cubes lose their orderly shape as they melt into water. Billowing smoke from a fire becomes increasingly disordered.

Ordered states become disordered, and that perhaps is the reason for the direction of the arrow of time. A measure of disorder tends to always increase in one direction of time. So maybe this is the answer. Maybe the arrow of time comes from the tendency of nature to evolve toward ever greater mess and disorder.

There is just one huge problem with this reasoning: because the laws of physics never distinguishes between the future and the past, entropy should increase not only toward the future, but must also increase towards the past. And this makes no sense at all. That is like saying that entropy must increase in either direction that we look.

When we look backwards in time and it should increase, and we look forwards in time, it should increase. That would mean the pages of your book in the past would be disordered and magically come together to form the neat, ordered book in your hands. But when is the last time that you saw something like this happen? Never!

How could our everyday experience in life be so at odds with the laws of math and physics? There is a big piece of the puzzle that is still missing and we do not understand at all. If we are sure that the past had to be ordered, that everything tends to go towards disorder as the equations of entropy says, then there something else besides the laws of physics that might explain this?

Think about hitting a baseball. The laws of physics can help us predict where it will land:

$$F = m * d^2x/dt^2$$

But those laws are not the only things that we need. We also need the *initial conditions*, like how hard the baseball was hit.

If the laws of physics also cannot give us an explanation for the arrow of time, then we need to look to our past for the *initial conditions* of the universe. That brings our attention back to the Big Bang. If the history of the universe is like a movie, and we ran this movie backwards, we would see an increase in order the further back in time we go.

Gradually to today's universe, with so many billions of galaxies clumped here and there, would turn back into clouds of gas and dust as everything contracts. As the clouds of gas and dust compact closer and closer to each other, so that if we get far enough into the past, they are squeezed into a smaller and smaller volume. We have found the answer. If the start of the Big Bang represents all of space at each moment of time, then we can see that there is not any more space-time before that single event. So the answer for the source of order, of low entropy, must be the beginning of the universe: the *Big Bang*.

The Big Bang

The Big Bang is a highly ordered state. It is the most ordered event in all of history and physics. And so, everything that came after that event has been an increase in disorder.

What the Big Bang gives us is a reason why the universe looks much different when we look backwards in time versus forward in time. Moreover, when we go back to early times, the universe looked not just different from today, but very highly ordered. Why was the entropy so low? We do not know. But at least we know for certain that there was point that the universe began in when the entropy was very low.

So the Big Bang is what set the *arrow of time* on its path. We can picture this as something like a wind-up clock. Just as the stored energy of a tightly coiled spring in the clock is released, and it unwinds, our entire universe has been unwinding since the Big Bang, and becoming ever more disordered.

The universe started in a very orderly state, and that is ultimately responsible for the fact that time have a direction. We do not know why the universe started in a very highly ordered state, but the fact that it did means that every time a glass of water shatters, it is actually carrying forward something that was set in motion billions of years ago.

For example, Quantum entanglement theory says that particles can influence each other even if they are separated by time and space in great distances, like billions of miles or maybe the entire length of the universe

So the glass breaks but does not un-shatter because it is following a natural drive from order to disorder that began with the Big Bang.

Everything that we see around us, all the changes, from the formation of the planets, to the stars, to our lives, is all epiphenomena, surfers riding the wave of time of increasing disorganization in the universe that defines the difference between the past and the future.

So it appears that the Big Bang may have stamped the arrow of time on our world and universe, and everything that has happened since may simply be the drive toward ever greater disorder that began with the Big Bang event, 13.7 billion years ago.

If time had a beginning and disorder is always increasing, does that mean that time has an end and will end?

The End of Time

What will the universe be like in the far, future? Recent discoveries are shedding new answers and light on this question.

The explosive force of the Big Bang sent space moving and hurtling outward so fast. And as a result, the universe is still expanding today. Until recently, most physicists thought that expansion must be slowing down by now. They thought of space filled with galaxies, and like a car traveling down a highway. If the driver takes his foot off the gas, the car would gradually slow down.

So similarly, they physicists thought that the universe was expanding, but with time, at a slower and slower rate. But surprisingly, astronomers found the expansion of the universe is not slowing down at all. It is even accelerating. It is as if someone is not taking their foot off the gas pedal, but stepping on it more, causing a turbo booster to kick in. And that is making the expansion of the universe speed up so much.

The expansion will keep accelerating in the future, not slow down at all. This is against everything we thought would happen. This has very strange effects on the future. Since the expansion of our universe is accelerating at a faster rate, in the far future, after 100 billion years or so, all of the distant galaxies will have hurtled out of sight from Earth.

And it will appear as if our galaxy were in the middle of emptiness or a black hole. Our descendants will be at a terrible big loss.

Light from distant galaxies has to travel so far to reach Earth. So when we look at galaxies, we are actually looking back in time. So in the far future, when those distant galaxies are so far away, and no longer visible, humans will find that the past, in cosmic terms, is out of reach.

And as for the end of time, one theory suggest that eventually, black holes will fill and dominate the cosmos. And with time, they too will evaporate & vanish, leaving nothing but random particles drifting through space.

In a far distant time and future when everything has decayed and everything is just sort of smoothed out, there is no change. And without any change, we will not have a notion of the passage of time. Because if you do not have events happening, then it is impossible to see how you would even imagine that there was time. You will not even know which direction of time is forward and which is backward. So time itself will one day lose its meaning.

About 355 years ago, Isaac Newton, who was of the first physicist to think about time scientifically, wrote that he did not need to define time because it is something "*well-known to all*."

But in trying to test our experience of time with the true nature of time, we have been forced to challenge some of our most deeply held thoughts and beliefs.

We now know that in every event that goes from order to disorder, there is an intimate entanglement at play. There is a link to the Big Bang itself, giving us the arrow of time.

The common-sense notion that one true time rules the universe has given way to a new picture in which time is different for each and every one of us. And the flow of time, which seems to us as real as the flow of water in a river, may be nothing more than just an illusion. Past, present and future may all simply exist on equal footing.

Our everyday experience of time will always exert a powerful and great influence. We will continue to imagine and think that time is universal, that the past is gone forever, and that the future is yet to be. But because of our scientific discoveries and understanding, we can now look beyond our experience and recognize that we are truly part of a far richer and far stranger reality.

Why does time go forwards, not backwards?

The arrow of time began its journey at the Big Bang, and when the Universe eventually dies there will be no more future and no past. In the meantime, what is it that drives time ever onward?

When Isaac Newton published his famous Principia in 1687, his three elegant laws of motion solved a lot of problems. Without them, we couldn't have landed people on the Moon 282 years later. But these laws brought to physics a new problem, which wasn't fully appreciated until centuries after Newton and still nags at cosmologists today.The issue is that Newton's laws work about twice as well as we might expect them to. They describe the world we move through every day – the world of people, the hands that move around a clock and even the apocryphal fall of certain apples – but they also account perfectly well for a world in which people walk backwards, clocks tick back afternoon to morning, and fruit soars up from the ground to its tree-branch.

"The interesting feature of Newton's laws, which wasn't appreciated till much later, is that they don't distinguish between the past and the future," says the theoretical physicist and philosopher Sean Carroll, who discusses the nature of time in his latest book The Biggest Ideas in the Universe.

"But the directionality to time is its most obvious feature, right? I have photographs of the past, I don't have any photographs of the future."

The problem is not confined to the centuries-old theories of Newton. Virtually all of the cornerstone theories of physics since then have worked just as well going forward in time as they do backwards, says physicist Carlo Rovelli of the Centre for Theoretical Physics in Marseille, France, and the author of books including The Order of Time.

"Starting from Newton, and then Maxwell's theory of electromagnetism, then Einstein's work, and then quantum mechanics, quantum field theory, general relativity, and even quantum gravity — there is no distinction between past and future," Rovelli says. "Which came as a surprise, because the distinction is so evident to all of us. If you make a movie, it's obvious which way is the future and which one is the past."

How does a clear direction of time emerge from these descriptions of the Universe, which all lack their own arrow of time? As Marina Cortês, an astrophysicist at the University of Lisbon, puts it: "There's a lot of implications that start with taking seriously the question, 'Why does time pass?'"

Part of the answer lies at the Big Bang nearly 14 billion years ago. Another insight comes from the opposite extreme, in the Universe's eventual death. But before embarking on this epic journey back and forth along the timeline of the Universe, it's worth stopping off in 1865, just as the first truly time-directional law of physics came hurtling down the tracks of the Industrial Revolution.

Gathering steam

In the 19th Century, when coal was shoveled into furnaces to generate steam power, scientists and engineers hoping to develop better engines embraced a set of principles that described the relationship between heat, energy and motion. They became known as the laws of thermodynamics.

In Germany, 1865, the physicist Rudolf Clausius stated that heat cannot pass from a cold body to a hot one, if nothing else around them changes. Clausius came up with the concept he called "entropy" to measure this behaviour of heat – another way of saying heat never flows from a cold body to a hot one is to say "entropy only ever increases, never decreases."

As Rovelli stresses in The Order of Time, this is the only basic law of physics that can tell apart the past from the future.

A ball can roll down a hill or be kicked back to its summit, but heat can't flow from cold to hot.

To illustrate, Rovelli picks up his pen and drops it from one hand to the other. "The reason this stops in my hand is that it has some energy, and then the energy is turned into heat and it warms up my hand. And the friction stops the bouncing. Otherwise, if there was no heat, this would bounce forever, and I would not distinguish the past from the future."

So far, so straightforward. That is, until you start to consider what heat is on a molecular level.

The difference between hot things and cold things is how agitated their molecules are – in a hot steam engine, water molecules are very excited, careening around and colliding into each other rapidly. The very same water molecules are less agitated when they coalesce as condensation on a windowpane.

Here's the problem: when you zoom in to the level of, say, one water molecule colliding and bouncing off another, the arrow of time disappears. If you watched a microscopic video of that collision and then you rewound it, it wouldn't be obvious which way was forwards and which backwards.

At the very smallest scale, the phenomenon that produces heat – collisions of molecules – is time-symmetric. This means that the arrow of time from past to future only emerges when you take a step back from the microscopic world to the macroscopic – something first appreciated by the physicist-philosopher Ludwig Boltzmann.

"So the direction of time comes from the fact that we look at big things, we don't look at the details," says Rovelli. "From this step, from the fundamental microscopic vision of the world to the coarse-grained, the approximate description of the macroscopic world – this is where the direction of time comes in.

"It's not that the world is fundamentally oriented in space and time," Rovelli says. It's that when we look around, we see a direction in which medium-sized, everyday things have more entropy — the ripened apple fallen from the tree, the shuffled pack of cards.

While entropy does seem to be inextricably bound up with the arrow of time, it feels a bit surprising — perhaps even disconcerting — that the one law of physics that has a strong directionality of time built into it loses this directionality when you look at very small things.

"What is entropy?" Rovelli says. "Entropy is simply how much we're forgetting about the microphysics, how much we are forgetting about the molecules."

The beginning and the end

If there is an arrow of time, where did it come from in the first place?

"The answer is embedded in the beginning of the Universe," says Carroll. "The answer is because the Big Bang had low entropy. And still, 14 billion years later we are swimming in the aftermath of that tsunami that started near the Big Bang. That's why time has a direction for us."

The extraordinarily low entropy of the Universe at the Big Bang is both an answer and an enormous question. "The thing we understand the least about the nature of time, is why the Big Bang had low entropy, why the early Universe was like that," says Carroll. "And I think honestly, as a working cosmologist, I think that my fellow cosmologists have dropped the ball on this one. They don't really take that problem seriously enough."

Carroll published a paper in 2004 with his colleague Jennifer Chen, in which they aimed to explain why the Universe had such low entropy close to the Big Bang, rather than just assuming or accepting this was the case.

"There's plenty of loopholes in the theory, plenty of aspects of it that are not completely baked – but I also think it is by far the best theory on the market," says Carroll. "It doesn't cheat."

Other cosmologists agree that it is indeed time to turn serious thought to this problem of the Universe's low entropy origins. "The likelihood of our current Universe having initial conditions of this kind, and not any other kind, is around one in 10 to the 10 to 124 (1:10^10^124)," says Cortês. (Another way of saying it is that the event had a probability of 0.00...01 – with 10^(10^124) zeroes omitted – a number so large it's awkward to express in conventional maths, Cortês notes.)

"I mean I could safely say, this is the largest number in modern physics, outside of philosophy or mathematics."

Simply taking such unlikely low-entropy origins as given is a grand case of "shoving the problem under the rug", Cortês says. "If physicists keep doing this, after a while it's going to be a very big pile under the rug. It's left to us cosmologists to explain why time only moves forward."

Even if we don't yet know why, the Universe's low entropy past is a plausible source of time's arrow. Like most things that have a beginning, the arrow will also have an end.

The first person to spot this was, once again, the Austrian physicist Ludwig Boltzmann. "Boltzmann thought, 'ah, entropy is growing in the Universe and maybe it's going to maximum at some point'," says Rovelli. At that point, heat would be evenly distributed throughout the Universe, no longer flowing from one place to another. There would be no energy available in a useful form for doing work – in other words, almost nothing interesting would be happening throughout the entire Universe. As astrophysicist Katie Mack describes it, "As that process continues, everything is decaying so much that all that's left is the waste heat of everything that ever existed in the Universe." This fate is known as the thermal death of the Universe, or heat death.

"Stars will stop burning, nothing will happen anymore. There will be nothing but small thermal fluctuations," says Rovelli. "Suppose this happens – and we don't know for certain if it's going to happen, but suppose it does – should we say that there is no time direction there? Of course there's no time direction, because every phenomenon that happened one way could also go one way or the other. Nothing will distinguish the two directions of time."

This is perhaps the strangest thing about the arrow of time: "It only lasts for a little while," says Carroll. It's very hard to picture what might happen if the arrow of time eventually vanishes. "When we think we produce heat in our neurons," says Rovelli.

"Thinking is a process in which the neuron needs entropy to work. Our sense of time passing is just what entropy does to our brain."

The arrow of time that arises from entropy brings us a long way closer to understanding why time only goes forward. But there may be more arrows of time than this one – in fact there is arguably an entire volley of arrows of time pointing from the past to the future. To understand these, we have to step from physics into philosophy.

Human time

The ways that we understand and experience time shouldn't be taken lightly, says Jenann Ismael, professor of philosophy at Columbia University, New York. If you think about your own experience of time, you may soon be able to recognise several of the psychological arrows that form a core part of human experience. One of these arrows is what Ismael terms "flow".

"If you look out at the world, you don't experience a purely static representation of the instantaneous state of the world," she says, like in a movie made up of a number of static frames every second. "We see directly that the world is changing."

This experience of the flow of time is built into our perception. "Vision isn't like a movie camera at all," says Ismael. "Actually what happens is your brain is collecting information over some temporal period. It's integrating that information so that at any given moment, what you're seeing is a computation that the brain has done. So that you not only see that things are moving, you see how fast they're moving, the direction in which they're moving. So the whole time, your brain is integrating information over temporal intervals and giving you the result. So you see time, in a way."

There's a second feature of time that Ismael distinguishes from flow, which she terms "passage".

The idea of passage is closely bound up with time-oriented experiences such as memory and anticipation. Take the example of a wedding, or any much-anticipated life event. Our experience of these moments has many layers — from the fractious planning stages, to the intensity of the day itself, to recollections that stay with us for years. There is a directionality to these different experiences: the way we anticipate an event in the future is fundamentally different from how we remember it when it's passed. "All of that is part of what I think of as the experience of passage, this idea that we experience every event as anticipated from the past, experienced in the present, remembered in retrospect," says Ismael. "It's kind of Proustian in its density."

These aspects of the directionality of psychological time – as well as many others, like the sense of openness we have about the future but not the past – could all trace their roots back to the arrow of time born of the Industrial Revolution.

"I think it does all come back to entropy," says Ismael. "I see no reason now to think that the kinds of arrows that are involved in human psychology are anything but ultimately rooted in the entropic arrow. But it's an empirical question. This project to understand human experience in relation to the entropic arrow, I've no reason to think it's going to fail."

That project is what Carroll hopes to do, taking several features of our experience of time and relating them back to entropy. His first target is causality, another element of the arrow of time, as causes happen before their effects.

To say the least, this project is a major undertaking for all physicists and philosophers involved. And still, lurking in the shadows behind all such efforts, there remains that nagging question about why entropy was so low in the earliest Universe. "I think we understand why we have this sense of flowing," says Rovelli. "We understand why the past seems fixed to us that the future seems open.

We understand why there are irreversible phenomena, and we can reduce all that to the second law of thermodynamics, to the rise of entropy.

"It's very much related to the fact that if we trace it back, back, back, to fact that the Universe started very small, in a very peculiar situation. Then somehow, it's falling down from that peculiar situation.

"But of course there's one question open, I mean, why? Why did it start in that particular way?"